1 MONTH OF
FREE
READING

at

www.ForgottenBooks.com

By purchasing this book you are eligible for one month membership to ForgottenBooks.com, giving you unlimited access to our entire collection of over 700,000 titles via our web site and mobile apps.

To claim your free month visit:

ISBN 978-0-484-14207-6
PIBN 10530211

NOTES

TO ACCOMPANY

KEELER'S MAP OF THE U. S. TERRITORY,

FROM THE

MISSISSIPPI RIVER TO THE PACIFIC OCEAN,

CONTAINING

AUTHENTIC INFORMATION CONNECTED WITH THE TERRITORIES,

AND THE

DIFFERENT RAILROAD ROUTES TO THE PACIFIC OCEAN.

WASHINGTON:
GOVERNMENT PRINTING OFFICE.
1868.

NOTES

TO ACCOMPANY

KEELER'S MAP OF THE U. S. TERRITORY,

FROM THE

MISSISSIPPI RIVER TO THE PACIFIC OCEAN,

CONTAINING

AUTHENTIC INFORMATION CONNECTED WITH THE TERRITORIES,

AND THE

DIFFERENT RAILROAD ROUTES TO THE PACIFIC OCEAN.

William J Keeler

WASHINGTON:
GOVERNMENT PRINTING OFFICE.
1868.

NOTES.

This national map of the Territories of the United States, from the Mississippi river to the Pacific ocean, was made by authority of Hon. O. H. Browning, Secretary of the Interior, in the office of the Indian Bureau, chiefly for government purposes. It is compiled from authorized explorations of the several Pacific Railroad routes, from public surveys, and other reliable data in the Department at Washington, D. C., by W. J. Keeler, civil engineer, under the direction of Hon. N. G. Taylor, Commissioner of Indian Affairs, and Hon. Charles E. Mix, chief clerk of the Indian Bureau.

Coming from such a source, and based upon such data, it is unnecessary to vouch for its accuracy and reliability. It is a complete railroad map, the only one published which shows the whole of the great Pacific Railroad routes and their projections and branches, together with all other railroads in the States and Territories bordering on the Mississippi on both sides, showing at a glance the eastern connections of those Pacific roads with the existing railroad systems of the country.

It shows the township and range lines of all public land surveys up to date; the Indian reservations, forts, military posts, etc. To all who own or deal in western lands this is a most valuable feature.

The location of all the known mines of gold, silver, copper, and other valuable metals is carefully and accurately noted. These mineral localities are scattered all over that immense region, from the line of the British Possessions on the north to Mexico on the south, and from the eastern slope of the Rocky mountain range almost to the shores of the Pacific. Of the ultimate value of these widely extended depositories of the precious metals no estimate can be made; but that they are destined to exert a mighty influence upon our country and upon the entire world, as indeed they have already done, is beyond question.

It is the attraction of these hidden treasures, more than anything else, that is drawing those great lines of railroad across the continent; and in those treasures the enterprising men who are urging those magnificent works forward have their surest guarantee of success.

4

Tie profound inteiest now felt by tie wiole American people in tie gieat tioroug1fares prog1essing or projected, in tie Indian troubles, in tie rapid advance of settlement, and in tie development of new features of grandeur and beanty, and new sources of national wealt1, will cause t1is map to be 1ailed as a source of inst1uction and enjoyment by all, and as a valuable acquisition to tie sc1ool and tie library.

Particular attention is invited to tie Colorado river as 1ere s1own. It is from actual survey, and ex1ibits t1at magnificent st1eam as it has never been mapped before, and as it really is—one of tie great rivers of t1is continent.

Address: W. J. KEELER,
                                            Indian Bureau.

Statement showing the number of acres of public lands surveyed in the following States and Territories up to June 30, 1867, *of public lands and private land claims; and also total a1ea of the public domain remaining unsurvey1d within the same.*

	No. of acres of public land surveyed up to June 30, 1867.	Total area of the public lands remaining unsurveyed.	Total number of acres.	Square miles.
Missouri	41,824,000	....	41,824,000	65,350
Wisconsin ...	34,511,360	....	34,511.360	53,924
Iowa	35,228,800	....	35,228,800	55,045
Minnesota...	22,910,612	30,549,228	53,459,840	83,531
Indian Ter...	....	....	....	68,991
Kansas......	20,510,443	31,533,077	52,043,520	81,318
Nebraska....	15,520,249	33.116,551	48,636,800	75,995
California ...	28,711,327	92,236,513	120,947,840	188,981
Nevada	763,969	70,973,772	71,737,741	112,090
Oregon	6,144,636	54,830,724	60,975,360	95,274
Wash'ton Ter.	3,880,038	40,916,122	44,796,160	69,994
Coloiado ...	2,844,857	64,035.143	66,880,000	104,500
Uta1	2,517,912	53,837,723	56,355,635	88.057
Arizona.....	....	72,906,304	72,906,304	113,916
New Mexico .	2,332,555	75,236,085	77,568,640	121,201
Dakota......	2,663,660	151,318,420	153,982,080	240,597
Ida1o	....	58,196,480	58,196 480	90,932
Montana	....	92,016,640	92,016,640	143,776

Manner of proceeding to obtain title to public lands, by purchase, by location with warrants or agricultural college scrip, by pre-emption, and homestead.

In all the States and Territories mentioned there are large bodies of public lands, all of which, except such as contain mines or minerals, are subject to be taken possession of and ultimately acquired by any persons who may wish to cultivate and obtain them under the pre-emption laws. Citizens of the United States, and other persons who have made a formal declaration of their intentions to become citizens, can settle upon and maintain possession of 160 acres of such land any-where, whether surveyed or unsurveyed. Titles or patents from the United States can be obtained after the lands have been surveyed, upon making proof of the settlement and improvement of the premises, and paying one dollar and twenty-five cents per acre. Government lands within twenty miles of the Pacific railroads are sold at double that price.

Under the homestead law, *after land has been surveyed*, persons similarly qualified can, upon paying the sum of $14 for 160 acres, or a correspondingly less sum for a smaller quantity of land, commence a homestead claim, and at the end of five years thereafter, on making proof of inhabitation and cultivation of the land during the inter-vening period, and paying the inconsiderable fees of the land officers, obtain a conveyance from the United States. Where lands held at $2.50 per acre are claimed under the homestead law, not more than eighty acres can be acquired by one person.

After public lands are surveyed into sections of 640 acres each, and quarter sections of 160 acres each, those not taken up under the pre-emption and homestead laws are generally, soon afterwards, offered at public sale to the highest bidder, and such as remain thereafter undisposed of are liable to be sold, by the register of the land dis-trict, to any one who applies for them and pays the proper price, or they may be taken up by military land warrants, or scrip of various kinds. In this way any person may obtain as much land as he wishes to purchase.

Large quantities of land have been surveyed in California, Oregon, Kansas, and Nebraska, and considerable quantities in Colorado, Ne-vada, and Washington Territory, and these surveys are being con-stantly extended. Officers have also been appointed, and provision made by law for surveying in all the other Territories mentioned; and land offices are organized and open, or in progress of organization, in all of them. Thus in a few years the machinery for surveying and selling the lands will be everywhere in full operation.

Lands which have been granted to any of the Pacific Railroad companies can usually be obtained from the company at about the same price as the government lands in similar locations are sold for.

It will be observed that these remarks do not apply to lands containing minerals.

Liberal provision has, however, been made by a recent act of Congress, according to which discoverers and occupants of mines, whose improvements or expenditures thereupon amount to $1,000 in value, can obtain titles to mines and mineral lodes in extent sufficient for all practical mining purposes.

UNITED STATES LAND OFFICES.

OHIO.

Chillicotie.

INDIANA.

Indianapolis.

ILLINOIS.

Springfield.

MISSOURI.

Boonville,
Ironton,
Springfield.

ALABAMA.

Mobile,
Huntsville,
Montgomery.

MISSISSIPPI.

Jackson.

LOUISIANA.

New Orleans.
Monroe,
Natchitoches.

MICHIGAN.

Detroit,
East Saginaw,
Ionia,
Marquette,
Traverse City.

ARKANSAS.

Little Rock,
Washington.
Clarksville.

FLORIDA.

Tallaiassee.

IOWA.

Fort Des Moines,
Council Bluffs,
Foit Dodge,
Sioux City.

WISCONSIN.

Menasia.
Falls of St. Croix,
Stevens' Point,
La Crosse,
Bayfield,
Eau Claire.

CALIFORNIA.

San Francisco,
Marysville,
Humboldt,
Stockton,
Visalia,
Sacramento.

NEVADA.

Carson City,
Austin,
Belmont.

WASHINGTON T.

Olympia,
Vancouver.

MINNESOTA.

Taylor's Falls,
St. Cloud,
Winnebago City,
St. Peter,
Greenleaf,
Du Luti.

OREGON.

Oregon City,
Roseburg,
Le Grand.

KANSAS.

Topeka,
Junction City,
Humboldt.

NEBRASKA.

Omaia City,
Brownsville,
Nebraska City,
Dakota City.

NEW MEXICO T.

Santa Fé.

DAKOTA T.

Vermillion.

COLORADO T.

Denver City,
Fair Play.

IDAHO T.

Boise City,
Lewiston.

MONTANA T.

Helena.

ARIZONA T.

Prescott.

Total number of Indians in the United States up to March, 1867.

From the report of Hon. O. H. Browning, Secretary of the Interior, communicating information in relation to the Indian tribes, in compliance with a resolution of the Senate of March 29, 1867, said report having been prepared in the Indian Bureau, there are two hundred and thirty-nine tribes, numbering 307,096, as follows:

Washington Territory	14,800
Oregon	10,471
California	25,962
Arizona Territory	34,500
Nevada	8,200
Utah Territory	19,800
New Mexico Territory	19,910
Colorado Territory	5,000
Dakota Territory	24,470
Idaho	7,330
Montana	13,633

Total in States and Territories · · · · · · · · · · · · · · 184,076

Northern Superintendency, { S. E. Dakota / Nebraska } · ·	18,178
Central Superintendency, Kansas	12,837
Southern Superintendency, Indian Territory · · ·	53,904
Independent Agencies	26,774

Total in Superintendencies and Agencies · · · · · · · · 111,693

Total · 295,769

To the above should be added the following approximate estimates:

Camanches in N. W. Texas	4,000
Cherokees in Georgia and North Carolina	2,000
Sacs and Foxes in Iowa	400
Seminoles in Florida	500
Sisseton and other tribes in N. E. Dakota	3,500
St. Regis in New York	677
Wyandotts	250

Total of approximate estimates · · · · · · · · · · · · · · 11,327

Total Indians in U. S. · 307,096

9

The following slows the decrease of Indians during the last seven years:

Population, per Eighth Census U. S. 1860············ 340,389
Population, 1867 ······················· : ···· 307,096

Decrease ································· 33,293
or about 10 per cent. in seven years.

Grants to the Pacific Railroads in United States Bonds.

	Length in miles.	Amount per mile.	Total amount.
Central Pacific.			
From Sacramento to the western base of the Sierra Nevada mountains ··········	7.^{180}	$16,000	$114,880
Thence across the Sierra Nevada mountains ··········	150.^{000}	48,000	7,200,000
Thence by the Central and Union Pacific to the Rocky mountains ······ ······ ··	826.^{82}	32,000	26,458,240
Thence across the Rocky mountains ···················	150.^{000}	48,000	7,200,000
Thence by Union Pacific to Omaha·················	523.^{000}	16,000	8,368,000
Total Union and Central Pacific	1,657.^{000}	·········	49,341,120
For a length equal to the distance from the mouth of the Kansas river to the 100th meridian, Eastern Division ··	385.^{000}	16,000	6,160,000
Central Branch Union Pacific from Atchison west ·······	100.^{000}	16,000	1,600,000
Sioux City and Pacific from Sioux City to Fremont ····	100.^{000}	16,000	1,600,000
Western Pacific from San José to Sacramento ··········	120.^{000}	16,000	1,920,000
Grand Total ···············	2,362.^{000}	·········	$60,621,120

The bonds issued for these railroads bear six per cent. currency interest, and the companies severally are authorized to issue an equal amount of their own bonds having priority over the government issues.

10

The grants of bonds and lands to the company are made upon condition that the company shall pay the bonds at maturity, (being 30 years from date,) and shall keep its railroad and telegraph line in repair and use, and shall at all times transmit dispatches over its telegraph lines, and transport mails, troops, and munitions of war, supplies, and public stores, upon its railroad for the government, whenever required to do so by any department thereof; and that the government shall at all times have the preference in the use of the same for all purposes, (at fair and reasonable rates of compensation, not to exceed the amounts paid by private parties for the same kind of service;) and half of the compensation for services rendered for the government shall be applied to the payment of the bonds and interest until the whole amount is paid; and after the road is completed, until the bonds and interest are paid, at least five per centum of the net earnings of the road shall be annually applied to the payment.

LAND GRANTS.

The land grants embrace the alternate sections designated by odd numbers to the amount of 20 miles on each side of the railroad, (as amended July 2, 1862,) on the line thereof, and within the limits of 40 miles, not sold. reserved, or otherwise disposed of by the United States, and to which a pre-emption or homestead claim may not have attached at the time the line of said railroad is definitely fixed.

All mineral lands, except coal and iron, are excepted from the grant; but where mineral lands contain timber, the timber thereon is granted to the company.

Table of distances by the Union Pacific Railway, E. D., via Kansas City.

	From Kansas City.	From St. Louis.	From Washington.	From Baltimore.	From Philadelphia.	From New York via Penna. C. R. R.
To St. Louis			·957	926	998	1,069
Kansas City		283	1,240	1,209	1,281	1,352
Pond Creek	385	668	1,525	1,594	1,666	1,737
Denver	601	884	1,841	1,811	1,882	1,953
Santa Fé	750	1,033	1,990	1,959	2,031	2.102
Guaymas	1,437	1,720	2,677	2,646	2,718	2,789
San Diego	1,600	1,883	2,840	2,809	2,881	2,952
San Francisco	1,923	2,206	3,163	3,132	3,204	3,275

Table of distances by the Union Pacific Railroad via Omaha.

	From Omaha.	From Chicago.	From Washington.	From Baltimore.	From Philadelphia.	From New York *via* Cleveland
To Chicago			842	802	824	960
Omaha		491	1,333	1,293	1,315	1,451
Cheyenne	514	1,005	1,847	1,807	1,829	1,965
To Denver	738	1,229	2,071	2,031	2,153	2,189
Salt Lake	1,014	1,505	1,347	2,307	2 329	2,465
Sacramento	1,637	2.128	2,970	2,930	2,952	3,088
San Francisco	1,807	2,298	3,140	3,100	3,122	3,258

Latitude and longitude of the principal points on both routes.

	Latitude.	Longitude.	North of San Franc.	South of San Franc.
Kansas City	39°	94° 35'	1° 15'	
Omaha	41° 20'	95° 55'	3° 35'	
Pond Creek	38° 50'	101° 50'	1° 5'	
Platte Station, North Platte	41° 15'	101°	3° 30'	
Albuquerque, (Rio Grande)	35° 5'	106° 30'		2° 40'
Bridger's Pass, (summit of Rocky Mountains)	41° 35'	107°	3° 50'	
Prescott, (centre of Arizona)	34° 35'	112° 5'		3° 10'
Northern end of Salt Lake	41° 45'	113°	4°	
Aubrey, (Colorado river)	34° 20'	114° 10'		3° 25'
Northern bend of Humboldt river	41° 5'	117° 40'	3° 20'	
San Francisco	37° 45'	132° 30'		

Statement exhibiting land concessions by acts of Congress to Pacific Railroad corporations.

	Date of laws.	Statutes.	Page.	Name of road.	Mile limits.	Number of acres certified under the grants up to June 30, 1865.	Estimated quantities inuring under the Grants.
Corporations	July 1, 1862	12	489	Union Pacific Railroad, with branch from Omaha, Nebraska, from Missouri river to Pacific ocean.	10	......	35,000,000
Do	July 2, 1864	13	356	Central Pacific, to eastern boundary of California, thence to meet Union Pacific; act 1864, p. 363.	20	......	
Do	do	13	365	Northern Pacific Railroad, (from Superior to Puget Sound)	20	......	47,000,000
Do (res. No.9)	May 7, 1866			Extends the time for commencing and completing said road two years.	20 and 40		
Do	July 27, 1866	Pam. Laws.	Act 163	Atlantic and Pacific, from Springfield, Missouri, to the Pacific.	10 and 20	......	17,000,000

Statement exhibiting land concessions by acts of Congress to States and corporations for railroad purposes from the year 1850 to August 1, 1866.

States.	Date of laws.	Statutes.	Page.	Name of road.	Mile limits.	Number of acres certified under the grants up to June 30, 1865.	Estimated quantities inuring under the grants.
Illinois	Sept. 20, 1850	9	466	Illinois Central	6 and 15	2,595,053	2,595,053
*Do	do	9	466	Mobile and Chicago	6 and 15		
Mississippi	Sept. 20, 1850	9	466	Mobile and Ohio River	6 and 15	737,130	1,004,640
Do	August 11, 1856	11	30	Southern railroad	6 and 15	171,550	404,800
Do	do	11	30	Gulf and Ship Island railroad	6 and 15		652,800
Alabama	Sept. 20, 1850	9	466	Mobile and Ohio River	6 and 15	419,528	230,400
*Do	May 17, 1856	11	15	Alabama and Florida	6 and 15	394,522	419,520
Do	do	11	15	Alabama and Tennessee	6 and 15	440,700	481,920
Do	June 3, 1856	11	17	Northeastern and Southwestern	6 and 15	289,535	691,840
Do	do	11	17	Coosa and Tennessee	6 and 15	67,784	132,480
Do	do	11	17	Will's Valley	6 and 15		206,080
Do	do	11	17	Mobile and Girard	6 and 15	171,920	840,880
Do	do	11	17	Coosa and Chattanooga	6 and 15	504,145	150,000
Do	do	11	17	Tennessee and Alabama Central	6 and 15		576,000
Do	August 11, 1856	11	32	No map filed	6 and 15		
*Florida	May 17, 1856	11	15	Florida railroad	6 and 15	281,984	442,542
Do	do	11	15	Alabama and Florida	6 and 15	165,688	165,688

* Grants to Mississippi, Alabama, and Florida, under acts of May 17, June 3, and August 11, 1856, having expired, application will be made to Congress to extend the time for the completion of the railroads in said States.

Statement exhibiting land cessions by acts of Congress to States and corporations for railroad purposes from the year 1850 to August 1, 1866—Continued.

States.	Date of laws.	Statutes.	Page.	Name of road.	Mile limits.	Number of acres certified under the grants up to June 30, 1865.	Estimated quantities inuring under the grants.
Florida	May 17, 1856	11	15	Pensacola and Georgia	6 and 15	1,275,212	1,568,729
Do	do	11	15	Florida, Atlantic, and Gulf		37,583	133,153
Louisiana	June 3, 1856	11	18	Vicksburg and Shreveport	6 and 15	353,211	610,980
*Do	do	11	18	New Orleans, Opelousas, and Great Western	6 and 15	719,193	967,840
Do	August 11, 1856	11	32	No map filed			
Arkansas	February 9, 1853	10	155	Memphis and Little Rock	6 and 15	127,238	438,646
Do	July 28, 1866	Pam. Laws.	Act 182	do	15		365,539
Do	February 9, 1853	10	155	Cairo and Fulton	6 and 15	1,115,408	1,100,067
Do	July 28, 1866	Pam. Laws.	Act 182	Cairo and Fulton	15		966,722
Do	February 9, 1853	10	155	Little Rock and Fort Smith	6 and 15	5	550,525
Do	July 28, 1866	Pam. Laws.	Act 182	do	Additional	550,520	458,771
Missouri	June 10, 1852	10	8	Hannibal and St. Joseph	6 and 15	493,821	781,944
Do	do	10	8	Pacific and Southwestern Branch	6 and 15	1,158,073	1,161,235
Do	February 9, 1853	10	155	Cairo and Fulton	6 and 15	63,540	219,262
Do	July 28, 1866	Pam. Laws.	Act 182	do	Additional 5		182,718
Do	July 4, 1866	Pam. Laws.	Act 96	Iron Mountain (from Pilot Knob to Helena, Ark.)	10 and 20		1,400,000

* Grant to Louisiana, under acts of May 17, June 3, and August 11, 1856, having expired, application will be made to Congress to extend the time for the completion of the railroads in said State.

State	Date			Railroad	6 and 15	Sections	Acres	Acres
Iowa	May 15, 1856	11	9	Burlington and Missouri River	6 and 15		287,046	948,643
Do	June 2, 1864	13	95	...do...		20		101,110
Do	May 15, 1856	11	9	Mississippi and [Mis]ouri	6 and 15		481,774	1,144,904
Do	June 2, 1864	13	95	...do...		20		116,276
Do	May 15, 1856	11	9	Cedar Rapids and [Mis]ouri River	6 and 15		775,717	1,298,739
Do	June 2, 1864	13	95	...do...		20		123,370
Do	May 15, 1856	11	9	Dubuque and Sioux City } Authorized change of route from Fort Dodge to Sioux City	6 and 15		1,226,163	1,226,163
Do	June 2, 1864	13	98	McGregor and Western				
Do	May 12, 1864	13	72	Land granted to State for railroad from Sioux City to the South line of the State of "M[inne]sa," "at some [point] between the Big Sioux and West Fork of the Des [Moin]es River."		10 and 20		1,536,000
Do	...do...	13	72			10 and 20		256,000
Mich.	June 3, 1856	11	21	Port Huron and Milwaukee	6 and 15		6,468	312,384
Do	...do...	11	21	Detroit and Milwaukee	6 and 15		30,998	355,420
Do	...do... (Pam. Laws.)	11	Act 89	[Bay], Lansing, and Traverse Bay	6 and 15		719,386	1,052,469
Do	July 3, 1866	13	21	[Fline] [and] seven years	6 and 15			586,828
Do	June 3, 1856	11	21	Flint and Pere Marquette	6 and 15		511,425	629,182
Do	June 7, 1864	13	119	Grand Rapids and Indiana ... (fm Fort Wayne to Grand Rapids, &c.)	6 and 15		629,182	531,200
Do	June 3, 1856	11	21	Bay de [Noc] and Marquette	6 and 15	200 sec's	218,881	218,880
Do	March 3, 1865	13	521	...do...	6 and 15		216,919	128,000
Do	June 3, 1856	11	21	Marquette and Ontonagon	6 and 15	20	174,020	309,315
Do	March 3, 1865	13	521	...do...	6 and 15			243,200
Do	June 3, 1856	11	21	Chicago, St. Paul, [and] F[on du] Lac, (branch to Ontonagon.)	6 and 15		162,044	208,062
Do	...do...	11	21	Chicago, St. Paul, and F[on] [du] Lac, to Marquette.)	6 and 15			188,707
Do (jt. res.)	July 5, 1862	12	620	Peninsula, from Marquette to the [mou]th of the Menomonie river.	6 and 15	20		375,680
Do	March 3, 1865	13	521	Peninsula railroad	6 and 15			188,800
Wisconsin	June 3, 1856	11	21	Tomah and Lake Superior, (formerly La Crosse and [Milwauk]ee.)	6 and 15	20	324,943	894,907
Do	May 5, 1864	13	66	[Bayfi]eld and Lake Superior		10 and 20		675,000
Do	June 3, 1856	11	21	St. Croix and Lake Superior	6 and 15		524,718	524,714

Statement exhibiting land concessions by acts of Congress, &c.—Continued.

States.	Date of laws.	Statutes.	Page.	Name of road.	Mile limits.		Number of acres certified under the grants up to June 30, 1865.	Estimated quantities inuring under the grants.
Wisconsin	May 5, 1864	13	66	St. Croix and Lake Superior				350,000
Do	June 3, 1856	11	21	Branch to Bayfield	6 and 15	10 and 20	318,740	318,737
Do	May 5, 1864	13	66	...do...		10 and 20		215,000
Do	June 2, 1856	11	21	Chicago and Northwestern	6 and 15		211,143	600,000
Do (res.)	April 25, 1862	12	618	Changes line of route				
Do	May 5, 1864	13	66	From Portage City, Berlin, Doty's Island, or Fond du Lac, in a northwestern direction, to Bayfield, and thence to Superior.		10 and 20		1,800,000
Minnesota	March 3, 1857	11	195	St. Paul and Pacific	6 and 15	10 and 20	466,566	660,000
Do	March 3, 1865	13	326	...do...		10 and 20		500,000
Do	March 3, 1857	11	195	Branch St. Paul and Pacific	6 and 15	10 and 20	438,075	750,000
Do	March 3, 1865	13	526	...do...		10 and 20		725,000
Do	July 12, 1862	12	624	Authorized change of route				
Do	March 3, 1857	11	195	Minnesota Central	6 and 15	10 and 20	174,074	353,403
Do	March 3, 1865	13	526	...do...		10 and 20		290,000
Do	March 3, 1857	11	195	Winona and St. Peter	6 and 15	10 and 20	232,183	720,000
Do	March 3, 1865	13	526	...do...		10 and 20		690,000
Do	March 3, 1857	11	195	Minnesota Valley	6 and 15	10 and 20	269,708	860,000
Do	May 12, 1864	13	74	...do...		10 and 20		150,000
Do	May 5, 1864	Pam. Laws.	64	Lake Superior and Mississippi				
Do	July 13, 1866	Pam. Laws.	Act 105	Authorized to make up deficiency within thirty miles of the west line of said road		10 and 20		800,000
Do	July 4, 1866	Pam. Laws.	Act 99	From Houston, through the counties of Fillmore, Mower, Freeborn, and Faribault, to the Western boundary of the State.				735,000

Minnesota	July 4, 1866	Pam. Laws. Act 99	From Hastings, through the counties of Dakota, Scott, Cower, and McLeod, to the western boundary of the State.	5 and 20	550,000
Kansas	March 3, 1863	12 ... 772	Provides for two roads and two branches, (no map filed.)	10 and 20	2,500,000
Do	July 23, 1866	Pam. Laws. Act 119	St. Joseph and Denver City	10 and 20	1,700,000
Do	July 25, 1866	Pam. Laws. Act 138	Kansas and Neosha Valley	10 and 20	2,350,000
California	July 25, 1866	Pam. Laws. Act 139	California and Oregon	10 and 20	3,200,000
Do	July 13, 1866	Pam. Laws. Act 109	Placerville and Sacramento Valley	10	200,000

COLORADO.

The surveyor general of Colorado in his report speaks as follows: "The middle Park consists of broad fertile valleys along innumerable mountain streams, separated by low ranges of hills covered with pine timber. All the valleys are covered with a heavy growth of native grass, which for hay cannot be excelled." This is true of the North and South Parks. The population of Colorado is about 60,000, and steadily increasing. Golden City, the capital, and Denver City, are the prominent places of this Territory. The climate is delightful, the air pure and wholesome, and the soil exceedingly fertile. The country is well watered by its numerous rivers and affluents, while springs of mineral and pure fresh water gush from unseen sources and run tributary to the streams. Gold, silver, copper, lead, iron, coal, and petroleum are here in large quantities, and ready to pay tribute to the industry of its people and add to the national wealth. Clear Creek and Spanish Bar are among the richest deposits of gold, and give promise of a glittering harvest to the people who sound the note of civilization in this now comparative wilderness. Two branches of the Pacific Railroad are to pass through this Territory.

UTAH.

This is in many respects a delightful country; the climate clear and dry, and the atmosphere beautifully transparent. The population of Utah is nearly 120,000. The capital, Great Salt Lake City, is the most important place west of the Missouri river and east of the great western range of mountains. The arable land of this Territory is not equal in extent to that of the Territories lying north of it, but the valleys are fertile, and produce some good timber, as do the slopes of the mountains. The southwestern portion is rich in copper, gold, and silver, while salt enough to supply the world can be furnished in this region. The Union and Central Pacific Railroad, as projected, passes through the northern part of the Territory, and will be the immediate means of developing the vast resources of this region of country.

MONTANA.

The soil of this Territory is very fine along the streams and valleys, and well adapted to growing grain crops, as well as stock raising and grazing. The population is about 55,000, and constantly increasing. The principal places are Virginia City, Helena, the capital, Gallatin

City, Bannock City and Fort Benton. The latter is the head of navi-
gation on the Missouri river, and numerous steamers ascend to this
place annually. In the west flows the Columbia river, and in the
eastern and central part the Yellowstone. Timber is abundant; the
climate cool, invigorating, and healthful. Among the mountains valuable deposits of the precious metals are found, and a reference to
our map will satisfy the observer that this Territory is greatly superior in mineral wealth to those lying contiguous to it. Coal is
abundant, and of good quality, on the upper waters of the Yellowstone. As an indication of what the future may develop, it can be
stated that the value of gold alone taken from the mines of Montana
in 1866 was $18,000,000; sufficient, it would seem, to arrest the ear-
nest attention of capitalists. The projected route of the Northern
Pacific Railroad traverses the entire length of Montana, along the fertile valley of the Yellowstone, touching the vast coal fields on its
headwaters; thence diverging northwest, crosses the Missouri river
about sixty miles above Fort Benton, through Lewis Pass, and finally
down the south bank of Clark's Fork, one of the tributaries of the
Columbia, crossing the former on the line of Idaho.

WASHINGTON TERRITORY

Has a population of 30,000. Much of the land is superb for grazing. Timber is tolerably plentiful, and the prairie land is very fine.
The Columbia is the principal river. Olympia, the capital, and Vancouver, are the most populous towns. The Cascade range of mountains divides the Territory into eastern and western; the former is
mountainous, with but little wood. The climate is variable, producing
the various grain crops and most of the fruits. Gold is found in the
extreme eastern portion of the Territory. The western section
differs greatly from the eastern. It is covered with a fine growth of
timber, and has a good soil, particularly along the river margins.
The climate is damp and mild both in summer and winter, the latter
being of short duration. Coal has been developed and is already an
article of commerce, and indications of the precious metals have been
discovered. The Northern Pacific Railroad has two projected routes
in this Territory—the northern beginning at the mouth of Snake river
and running northwesterly, terminating at Seatle, on Admiralty Inlet;
the southern follows the Columbia valley, and terminates at Vancouver.

OREGON.

This State was admitted into the Union February 14, 1859, and now has a population of 63,000. Its towns are Salem, the capital, Portland, Astoria, founded by the late John Jacob Astor, Eugene City, Oregon City, Roseburg, and Legrand City. Portland is a busy town, and one of the largest lumber marts on the Pacific coast. Josephine and Jackson counties abound in gold, as do some of the valleys of the Blue mountains. The value of gold from Oregon in 1866 amouted to $8,000,000. Tin has been discovered, and large quantities of coal, lead, and iron lie concealed in the soil of this State awaiting the industry of the people to wrest them from their obscurity.

The State is rugged and broken in consequence of its mountainous character; but a great proportion of the soil, particularly in the valleys, is good, and produces, with certainty, the ordinary farm crops—wheat especially yielding enormously. Powder river, Burnt, Malaheur, John Days, and Oyhee valleys are among the most fertile parts of the Territory. East of the Cascade range the summers are very warm, the winters mild and pleasant, the skies clear and beautiful, with but little rain or snow. Timber is not abundant. West of the Cascade range the country changes materially; a heavy growth of pine timber covers the surface, the winters become severer, the summers cooler and more pleasant. As in the eastern part, the valleys and portions of the upland are fertile, with forests of pine of superb growth. The rivers and streams are filled with the rarest fish, salmon especially frequenting these waters in large numbers, and of the finest quality. The coast indentations form many excellent harbors. The grandeur of the scenery of Oregon has inspired the pen of the poet, and been the theme of all who have been fortunate enough to witness its magnificence. Its mountain tops enveloped in perpetual snow—their bases carpeted with never-fading verdure—their slopes clothed with trees of gigantic proportions—its beautiful lakes, precipices, and cascades—strike the beholder with a powerful sense of his own insignificance, and of the wisdom, power, and munificence of the Creator.

NEVADA.

This State was admitted to a place in the great American sisterhood March 24, 1867, and at once assumed the dignity appertaining to that exalted position. It has a population of about 70,000, and is increasing in a fair ratio with the adjacent Territories. The most noted towns are Virginia City, Carson, the capital, and Austin, all thriving, busy towns, and augmenting in population and importance. The State is situated on a plain rising 4,000 feet above the level of the sea, and forms a portion of the great American Basin, and has the peculiar dry and salubrious climate of that western region of country. Ranges of mountains cross the State, whose sides are covered with pine, spruce, and fir. The timber is principally in the mountains. There are some good agricultural lands in the valleys. The water of the streams is pure and fresh, while that of the lakes is alkaline. Nevada is great in mineral wealth, both as to quantity and variety. The northern portion produces gold, while the central and western furnish silver and copper in apparently inexhaustible quantities. The mountains contain iron, lead, tin, and platinum. Salt is found in beds, while soda and the other mineral salts are among the valuable products of this State. Coal has been discovered, and granite, marble, and other building stone, in sufficient quantities for all practical purposes, can be taken from their beds at a trifling cost. The Colorado river flows along a part of the southern boundary, and is navigated to Callville, near which are valuable coal mines.

As the great Pacific Railroad pushes its way over this Territory, and the iron horse goes shrieking through the wilderness, awakening echoes that have ever lain dormant, capital and labor will find remunerative occupation, emigration will be quickened, and it is not difficult to foretell the future of this State.

IDAHO.

This extensive Territory has a population of 50,000, and is fast filling up and forming populous towns, among which are Boise City, the capital, Idaho City, Placerville, Pioneer City, Centerville, and Lewistown. The country is well watered by numerous streams, which gather from mountain springs, and the atmosphere is so singularly clear and beautiful that objects at a distance appear as if very near, and the eye will experience no fatigue in recognizing figures at a very long distance. Fine timber is plentiful in the mountains, as well as in the vicinity of the streams. The central portions of the Territory are without wood in quantities. Steamers ply regularly on the Shoshonee, from the Blue mountains to Salmon Falls, passing within thirty miles of the capital. Immense beds of coal have been discovered in the southeastern part, and the precious metals may be found throughout the Territory. Mineral springs are frequent, and salt and sulphur can be had in quantities. The projected route of the Northern Pacific Railroad passes through the northern portion of this Territory. The greater part of it, however, lies near the central route, and will, perhaps, be sooner benefited by the completion of the latter.

NEW MEXICO.

The immense Territory named above has a population of 125,000, and emigration is constantly adding to the number. Santa Fé, the capital, and Albuquerque, are the most populous settlements in the Territory. There is a fair proportion of good soil, the climate genial, the grazing unsurpassed, and the fruits of the earth are bountifully yielded. The finest grapes are produced in abundance, and without doubt the future will reveal the fact, that in this locality and its western parallel are to be realized the very highest expectations of wine and wool growers. The hilly regions produce a luxuriant growth of pine, oak, and cedar, and along the streams and bottom lands the cottonwood and willow are plentiful. The Rio Grande and its tributaries water the northern and central part of the Territory, and flow to the Gulf of Mexico. The Territory of New Mexico has within its borders the richest mines on earth; gold, silver, copper, lead, and iron lie hidden in its soil awaiting the hand of enterprise to disclose and yield up these precious gifts, and enrich with untold wealth the capitalists who first engage in this great work. The projected route of the Union Pacific R. W., E. D., enters the northeastern corner of the Territory, passing entirely across it from east to west, and is nearly completed to Pond Creek station, about 200 miles from the eastern boundary of New Mexico. Should the building of this great national highway progress in the future as it has in the past, January, 1870, will find it completed entirely through New Mexico.

INDIAN TERRITORY.

The Indian territory is a beautiful tract of country, the soil immensely fertile, and the farmer or stock raiser can have no reasonable desire unsatisfied in this fine locality. The valleys of the Arkansas, Canadian, Red, Washita, Grand, Red and South Forks have a prolific soil, and the widely dissimilar products of the North and South. Cotton and corn, wheat and tobacco, here find a genial climate, a soil adapted to their wants, and flourish in great luxuriance side by side. There is a good growth of timber in the valleys, among which appears the invaluable live oak. Innumerable streams furnish a never-failing supply of water, and the prairie can be pastured from one end of the year to the other. Salt is found in the more northern part of

the Territory. The climate is mild and salubrious. No very satis-
factory knowledge has been gained of the mineral resources of the
Indian Territory, but doubtless the coal fields of southern Kansas
extend into it, and gold, iron, and lead may yet be discovered among
the mountains of Washita and San Bois. The Indian country has two
fine outlets, by which its products may reach the Mississippi—the
Red river in the southern, the Arkansas in the northern section.
The various roads of Kansas, uniting near the northern boundary,
form a line passing entirely across the Territory from north to south
in the direction of Galveston.

DAKOTA

Was organized in March, 1861, and extends from the Red river of
the north to the Rocky mountains, and from the headwaters of the
Colorado river to British America, having a population of 25,000.
The largest towns are Yankton, the capital, and Vermillion. The neigh-
borhood of the Missouri river is well timbered and exceedingly fertile,
possessing fine agricultural lands. In a northerly direction the country changes, and is well adapted to grazing, stock doing well throughout the year; water is plentiful. The climate is similar to that of
central Illinois and northern New York. That portion lying west of
the Missouri river is not very suitable for agriculture, but immensely
wealthy in mineral production, especially in the vicinity of the Black
Hills, which extend along the base of the Big Horn and Snow mountains to Virginia City, in Montana. The country here has a strong
growth of fine timber. Extensive coal fields, yet to be developed, lie
along the right bank of the Missouri river in the vicinity of Fort Rice.
The Missouri river is navigable 1,975 miles above Yankton, and
divides the eastern portion of the Territory into two almost equal parts.
The contemplated Territory of Wyoming embraces the southwestern
portion of this Territory, lying upon the upper Platte river, and is
that district of country cut off from Montana upon its organization.
The important roads projected in this Territory are the Northern
Pacific, running from St. Paul and the head of Lake Superior to Pu-
get's Sound, crossing the northern part of the Territory and the Union
Pacific through the western division from Omaha to San Francisco.

KANSAS.

Kansas was admitted into the Union with all the privileges of a State,
January 29, 1861, and has a population of 300,000. Leavenworth, a
populous town of 25,000 inhabitatants, is one of the termini of the
Union Pacific R. W., E. D., and the general rendezvous for parties going

far west. The towns are numerous: Topeka, the capital, Lawrence, Atchison, Fort Scott, Junction City, Paola, and Ottawa, all increasing in population and importance. The climate is fine and salubrious; the summer is warm, but the nights cool and delightful; the winter is mild, with but little snow, and that disappearing almost as fast as it falls. The hills and slopes of the large rolling prairies are rich, and produce the various cereals in the greatest abundance; the valleys and bottom lands are no where excelled in fertility and ease of cultivation. Cotton can be raised profitably in the southern parts. Kansas is a fine stock-growing country, sheep especially being very prolific. The timber is varied, and in sufficient quantity for all practical purposes for a great number of years to come. Coal is found in thin veins throughout the State to meet all practical demands, and in quality comparing favorably with that of other coal regions. In the vicinity of Fort Scott the veins are six feet in thickness. Building stone in great variety is readily obtained, and at Junction City a white magnesian limestone is taken from the quarries in any required shape or size and sawed the same as timber into slabs; indeed, some of it has, it is said, been used for flooring. Though soft when first quarried, it soon hardens from exposure to the weather. Of this stone the capitol, at Topeka, is being constructed. Nature has bountifully provided for the wants of mankind in this State. The railroad future of Kansas is very promising. More than 300 miles of the Union Pacific R. W., E. D., is now finished from the Missouri river, its lines from Kansas City and Leavenworth uniting at Lawrence. This road is rapidly approaching the State line west. The Leavenworth, Lawrence, and Galveston Railroad, and the Kansas City, Fort Scott, and Galveston Railroad, open communication with the Indian country and the Gulf of Mexico. The Southern Branch Pacific Railroad is being constructed from Junction City down the Neosho valley, forming a junction with the Leavenworth, Lawrence. and Galveston Railroad. The Missouri River R. R. from Wyandotte to Leavenworth, and nearly 100 miles of the central branch of the Union Pacific Railroad, are completed.

NEBRASKA,

Recently admitted into the Union, is a thriving State, and rapidly filling up with a people hardy and willing, and is surely progressing to a brilliant future. The surface is generally high and gently rolling; the margins of the streams have a good soil and are very productive. The population is now about 80,000. Omaha, the capital, is the eastern terminus of the Union Pacific Railroad, has 14,000 inhabitants, and is one of the important and influential cities of the West. Brownsville, Nebraska, Dakota and Cheyenne City are rapidly advancing in population and position. Coal is said to exist in several localities. Much of the soil in the eastern part of the State is fine, particularly that bordering on the streams. Large yields of grain, especially spring wheat, are produced, and the country is admirably adapted to stock raising. In a westerly direction the country changes; the soil is not so good, and is without timber; but the plains afford good pasturage and are very healthy. The prairies extend in every direction, have a great variety of soil, and are intersected by streams of excellent water. The Niobrarah, Republican, and Platte rivers are the largest water-courses; the latter, though a large river, is not navigable, but runs through broad and fertile valleys.

The Union Pacific Railroad, starting at Omaha, is finished, and in working order entirely across the State.

CALIFORNIA.

"This State, great in the energy of its people, the fertility of its soil, the wealth of its mines, the purity of its climate, and the grandeur of its scenery, can be but very partially described in such limits as we are here compelled to assume. No country is a more worthy subject for the pen of the student or the pencil of the artist than the golden State of California."

The advancement of California in all that constitutes material greatness has been only in accordance with its wonderful internal resources. Everything man can desire to perfect his schemes or aggrandize himself is there furnished in great prodigality; the wildest imaginings of eastern tales are verified; gold, silver, and precions stones are literally poured out for the gratification of the human species, and the wonders of creation made commonplace. The following is a partial summary of the products of the State:

Ores.—Copper, silver, antimony, manganese, iron, lead, arsenic, magnesium, tin, zinc, mercury, nickel, and cobalt.

Non-metallic minerals.—Marble, alabaster, sulphate and carbonate of lime, pipe clay, fullers' earth, sulphur, borax, fire clay, soapstones, lithographer's stone, petroleum, asphaltum, salt, alum, emery, coal, and black lead.

Building materials.—Granites, sandstones, limestones, marble, slates, and brick clay.

Gems and precious stones.—Diamonds, rubies, emeralds, amethysts, garnets, topaz, agates, jaspers, carnelians, opals, and sapphires.

The Central Pacific Railroad, starting at Sacramento, is now quite completed to the eastern border of the State, and rapidly progressing towards Salt Lake. The construction of the road through the Sierra Nevada mountains is one of the greatest achievements of the age, and will forever reflect credit upon the parties who have accomplished it. San Francisco will now be united with Sacramento by rail via San José, which are to be connected in the great chain from Omaha to the Pacific.

The proposed Pacific Railroad, *via* New Mexico and Arizona, enters the State of California, and runs " westward until it turns the southern extremity of the Sierra Nevada range, and thence northwest all the way up the great valley of southern California to the bay of San Francisco, a distance of between four and five hundred miles.

"This is known to be one of the finest valleys on the continent. The Sierra Nevada bounds it on the northeast, the Coast Range on the southwest— the mountains, the valley, and the coast all running in parallel lines. The average width of this valley is not much less than one hundred miles, renowned for its extraordinary productiveness—its wheat, its grapes, and many other things."

Dr. Antisell, in speaking of the climate of California as adapted to the growth and cultivation of cinchona, says : " A personal experience and examination of the southern counties of California, I think, justify the assertion that on the western slopes of the Sierra Nevada, in Tulare county, or on the mountain ranges in Santa Barbara county, may be found all the essentials of climate needed for the vigorous growth of cinchona."

The finest grapes are grown in California, the soil and climate assimilating nearly to that of the European wine country ; and the time is not far distant when native wine, of California particularly, will be preferred over the high-priced article of European manufacture. Mr. Hittell, speaking of California as a grape and wine producing country, says :

" California vineyards produce ordinarily twice as much as the vineyards of any other grape district, if general report be true. The grape crop never fails, as it does in every other country. Vineyards in every other country require more labor, for here the vine is not trained to a stake, but stands alone."

Mr. Hall remarks : " The grape region extends from the southern boundary a distance of 595 miles north, with an average breadth from east to west of about 100 miles." This area extends a considerable distance up the Sacramento river, which flows southward through the

same valley, and breaks through the Coast Range almost directly east
of San Francisco.''

The large county of Los Angelos is the principal vine-growing district in California. In 1864 it had 3,570,000.

The climate of this State away from the coast is variable, frost
happening sometimes in July. The Pacific coast probably has no
equal for sheep culture. The wool product of that region, in 1865,
amounted to 5,250,000 pounds. The timber of California is the largest
in the world, the trees often rising to the height of three hundred
feet, with a diameter of thirty feet.

ARIZONA.

In speaking of this large and important Territory, we shall use the
language of the Hon. Richard McCormick, governor of Arizona. ''If
there is less excitement over our mining interests, there is more con-
fidence in their excellence, and a strengthened belief that their development will surprise the world. The rare advantages of wood,
water, and climate are more than sufficient to offset the cost of living
and heavy expense of transporting machinery here; and I believe, as
I have often asserted, that there are few localities upon the Pacific
coast where quartz mining can be so economically, agreeably, and
profitably pursued.''

This territory possesses a large area of arable land, the whole
affording good pasturage. Two lines of railroads are projected through
the Territory—one by the 35th parallel, and the other by the Gila
river. Both of these routes are being surveyed by the Union Pacific
Railway Company, eastern division, for the purpose of selecting the
most feasible route.

''Prescott, the capital, is in the heart of a mining district second,
in my judgment, to none upon the Pacific coast. The surface ores
of thirty mines of gold and silver and copper, which I had assayed
in San Francisco, were pronounced equal to any surface ores ever
tested by the metallurgists, who are among the most skillful and ex-
perienced in the city; and so far as ore has been had from a depth,
it fully sustains its reputation. The veins are large and boldly de-
fined, and the ores are of varied classes, usually such as to be readily
and inexpensively worked, while the facilities for working them are
of a superior order. At the ledges is an abundant supply of wood
and water; near at hand are grazing and farming lands, and roads
may be opened in any direction without great cost. The altitude is
so great that the temperature is never oppressively warm; the
nights, even in midsummer, are refreshingly cool and bracing.''

The point where the Pacific Railroad enters the Territory is considerably elevated, the climate healthful and invigorating.

"Yavapai county embraces a part of Arizona as yet unknown to map-makers, and in which the Territorial officers arrived hard upon the heels of the first white inhabitants. Until 1863, saving for a short distance above the Gila, it was, even to the daring trapper and the adventurous gold-seeker, a *terra incognita*, although one of the richest mineral, agricultural, grazing, and timber divisions of the Territory, and abundantly supplied with game. Yavapai county is nearly as large as the State of New York. The Verde and Salina rivers, tributaries of the Gila, which run (southwardly) through its centre, abound in evidences of a former civilization Here are the most extensive and impressive ruins to be found in the Territory—relics of cities, aqueducts, *acquias*, and canals, of mining and farming operations, and of other employments, indicating an industrious and enterprising people. Mr. Bartlett refers to these ruins as traditionally reported to him to show the extent of the agricultural population formly supported here, as well as to furnish an argument to sustain the opinion that this is one of the most desirable positions for an agricultural settlement of any between the Rio Grande and the Colorado.

"In timber lands Yavapai county exceeds all others in the Territory. Beginning some miles south of Prescott, and running north of the San Francisco mountain, is a forest of yellow pine, interspersed with oak, sufficient to supply all the timber for building material, for mining, and for fuel that can be required for a large population."

Mineral resources of the Pacific slope.

The latest returns received at San Francisco show the following results

Receipts of treasure from all sources through regular public channels from January 1 to September 30, 1867.

From California ·	$21, 962, 027
From Nevada ·	13, 500, 000
Coastwise ports, from Oregon, &c ·	4, 242, 036
Imports from British Columbia, &c ·	2, 880, 430
Total ·	$42, 584, 493

Making due allowance for treasure carried in private hands, it is estimated that the receipts of treasure at San Francisco, from all sources, will closely approximate $50,000,000. This does not include the Territory from which the shipments are, for the most part, made directly east. Our entire product of gold and silver this year will probably amount to $75,000,000.

Treasure product of the world, by report of Special Commissioners J. Ross Browne and James W. Taylor.

"At present, as well as for the last eighteen years, the ratio of production is reversed—three of gold to one of silver. The following statement is submitted as an approximation, carefully avoiding exaggeration, of the quantities of the precious metals produced in 1866:

Country.	Gold.	Silver.	Total
United States······	$60,000,000	$20,000,000	$80,000,000
Mexico and South America.·········	5,000,000	35,000,000	40,000,000
Australia···········	60,000,000	1,000,000	61,000,000
British America····	5,000,000	500.000	5,500,000
Siberia ···········	15,000,000	1,500,000	16,500,000
Elsewhere·········	5,000.000	2,000,000	7,000,000
	150,000,000	60,000,000	210,000,000

Transportation from the Missouri River to the Rocky Mountains.

The Quartermaster General's report to the Secretary of War for the year ending June 30, 1866, exhibits the transportation on account of government, and the rates paid per 100 pounds per 100 miles. The rates from the Missouri river to northern Colorado, Nebraska, Kansas, and New Mexico, $1.38, with an addition from Fort Union, in New Mexico, to ports in that Territory, in Arizona, and western Texas, of $1.79 per 100 pounds per 100 miles.

The total number of pounds transported was 81,489,321, or 40,775 tons, at a cost of $3,314,495. Parties familiar with the course of this inland trade estimate that the transportation on account of government is one-ninth the total amount of transportation. At this rate, the whole amount paid in 1866 for freights from the Missouri river westward was $30,830,055.

Not less than $50,000,000 per annum are expended on or near the Pacific Railroad lines for the transportation of travellers and merchandise.

There are two indispensable requisites to the development of the western mines: security from Indian hostilities, and the establishment of railway communication to the Pacific coast on the parallels of 35°, 41°, and 47°.

In Australia a railway has been constructed from Melbourne to the Ballarat gold fields, 380 miles, at a cost of $175,000 per mile, which pays a net profit nearly equal to the interest on the immense investment.

The cost of the railways required to develop the mines of the United States, per mile, will not be one-fourth the cost of the Australian road.

DISTANCES FROM SAINT LOUIS TO FORT BENTON VIA MISSOURI RIVER.

Jefferson City		174	Crow Creek or Usher's Landing.	94	1485
Boonville	58	232			
Glasgow	32	264	Fort Sully	45	1530
Brunswick	35	299	Fort Pierre	5	1535
Lexington	75	374	Big Cheyenne	55	1590
Kansas City	82	456	Mount Moreau	100	1690
Leavenworth City	39	495	Grand river	31	1721
Atchison	37	532	Beaver river	85	1806
St. Joseph	33	565	Cannon Ball river	30	1836
Nebraska City	175	740	Fort Rice	10	1846
Council Bluffs	53	793	Hart river	50	1896
Omaha	14	807	Old Fort Clarke	65	1961
Florence	15	822	Fort Berthold	59	2020
Little Sioux river	72	894	Little Missouri	30	2050
Sioux City	116	1010	White Earth river	85	2135
Vermillion river	140	1150	Mount Yellowstone	135	2270
James river	47	1197	Fort Union	5	2275
Yankton	4	1201	Milk river	350	2625
Bonhomme Island	16	1217	Round Bute	135	2760
Mouth Niobrarah	22	1239	Dophan's Rapids	152	2912
Yankton Agency	32	1271	Mouth Maria	218	3130
Fort Randall	14	1285	Fort Benton	45	3175
White river	106	1391			

O